# REPORT,

## ARTICLES OF ASSOCIATION AND BY-LAWS

OF THE

# Clifton Mining Company,

ORGANIZED OCTOBER 19th, 1853,

UNDER THE GENERAL MINING LAW OF THE

STATE OF MICHIGAN.

---

Office 35 Wall St., New-York.

---


NEW-YORK:
FRANCIS HART & COMPANY, PRINTERS,
63 CORTLANDT STREET.
1864.

# REPORT,

## ARTICLES OF ASSOCIATION AND BY-LAWS

OF THE

# CLIFTON MINING COMPANY,

ORGANIZED OCTOBER 19th, 1853,

UNDER THE GENERAL MINING LAW OF THE

STATE OF MICHIGAN.

---

Office 35 Wall St., New-York.

---

NEW-YORK:
FRANCIS HART & COMPANY, PRINTERS,
63 CORTLANDT STREET.
1864.

THE

# CLIFTON MINING COMPANY,

INCORPORATED UNDER THE GENERAL MINING LAW OF THE STATE OF MICHIGAN.

---

CAPITAL STOCK,

## FIVE HUNDRED THOUSAND DOLLARS,

In 20,000 Shares of $25 each.

---

## Officers.

HENRY G. DUNNEL, *President.*

---

### Directors.

H. G. DUNNEL, W. O. EWEN,
D. C. LAWRENCE, W. H. GRENELLE,
F. A. ARTAULT, P. C. BLANCAN,
GEORGE T FROST, *of Detroit, Mich.*

---

### Secretary and Treasurer.

P. C. BLANCAN.

FEBRUARY, 1864.

---

**Office 35 Wall Street, New-York.**

# PROSPECTUS

OF THE

# CLIFTON MINING COMPANY.

---

*The Clifton Mining Company* was organized at Detroit, Mich., in the year 1853, under the General Mining Law of Michigan.

The land of the Company consists of three quarter sections, viz: the southwest one-quarter of section 10, the northeast one-quarter of section 4, and the southwest one-quarter of section 15, in town 49, north of range 41 west, in the Ontonagon mining district of Lake Superior, Michigan, and includes 480 acres, 320 acres of which are mineral lands, and 160 acres agricultural.

The work was begun at the mine in November, 1853, under the superintendence of Capt. Angus, with the view of ascertaining the best bearing veins, before proceeding to heavy mining. The ground mined upon was on the south declivity of the trap range, between the *Norwich* and *Sharon* mines. The *Norwich* was then mining with full evidence of success, and the *Sharon* was opening very strong veins. The ground occupying the same position on the range, it was assured it would carry the same veins, and be correspondingly valuable, and this view of its value the exploration made seemed to justify.

There are 4 veins, but upon no one of them has there been sufficient work done to prove fully their value. They are, however, rich in copper. The discovery by the adit, which has been extended 148 feet in the rock, of the south vein, is highly important as identifying the position with the same vein of the mining range of the *Sharon Company,* and the proof of the vein, which this Company has made, is highly encouraging for the *Clifton.*

The amount of shafting done is 123 feet, and amount of drifting, 262 feet.

There were 5 buildings erected upon the property.

In consequence of the monetary difficulties of the country in 1855, the operations of the mine could not be continued.

The reproduction of the pamphlet published in 1855 by the Clifton Company, with map, may be interesting to the Stockholders, and are submitted herewith.

New-York, January 1st, 1864.

# REPORT OF DIRECTORS.

OFFICE OF CLIFTON MINING COMPANY,
*Detroit, October*, 1855.

The Directors of the Clifton Mining Company, in pursuance of a resolution adopted at a meeting of the Stockholders, on the 15th June instant, submit the following exhibit of the affairs of the Company, and of the condition and prospects of its mining interest.

The statement of the Treasurer, hereto annexed, shows the receipts and expenditures of the company from its organization; and the report of F. G. White Esq., Superintendent of the mine, presents in detail the operations at the mine. This report, read in connection with the map and diagram accompanying it, will, we trust, be very satisfactory. There are annexed also the Articles of Association and By-Laws of the Company, and the several acts of the Legislature of the State, showing the manner of organizing mining corporations, and the rights and interests of parties under them. This being the first printed report made to the stockholders, since the organization of the Company, it was deemed of interest to include these papers.

The lands of the Company consist of three quarter sections, viz.: the S. W. ¼ of Sec. 10, the N. E. ¼ of Sec. 4, and the S. W. ¼ of Sec. 15, in Town 49, north of Range 41 West, in the Ontonagon Mining district of Lake Superior, Michigan, and include 480 acres, 320 acres of which are mineral lands, and 160 acres agricultural.

The first above quarter, containing 160 acres, is the land upon which the mining operations of the Company have been conducted. It was purchased from Government, after a careful examination for mining purposes, by Capt. J. D. ANGUS, his patent bearing date Dec. 1, 1851, and was conveyed by him by warranty bearing date
1853, to CHAS. RICHMOND Esq., in trust, and by the latter, by like instrument, to the Company. The other quarters of land were entered

and purchased from Government by H. N. WALKER Esq., from whom the title passes by warranty to the Company.

The ground which the Company are mining lies upon the south declivity of the Trap Range, between the Norwich and Sharon mines. When purchased, the Norwich Company were mining with full evidence of success, and the Sharon were opening very strong veins Our ground lying between these mines and occupying the same position on the range, we were assured it would carry the same veins, and be correspondingly valuable, and this view of its value the exploration made seemed to justify.

Capt. ANGUS, on account of his greater familiarity with the location, was engaged to commence the improvements and work of mining. This was done on an economical scale, commencing in the fall of 1853, immediately following the organization of the Company. Substantial improvements, suited to the first work of mining, were made. and the work of exploring and mining has been economically and steadily prosecuted to the present time. The services of Capt. ANGUS were continued to the spring of 1854. Mr. F. G. WHITE has been in charge since that time, and it affords us pleasure to again refer to his report for the purpose of expressing our satisfaction with the manner in which he has managed affairs at the mine.

It has been our object to ascertain our best bearing veins, before proceeding to heavy mining. From the position of our ground, we expected to find the veins which the Sharon and Norwich Mining Companies are so successfully mining. The developments have clearly established that our south vein or vein No. 1, is the Sharon south vein, and we shall pursue the work on this vein with great confidence, from the rich proof *it* is making *at that* mine. That the Norwich vein also crosses this tract, there can be but very little doubt —it is more than probable even that it is the *above* vein—and that this ground will prove, as it is opened, very valuable in copper, the Directors and your Superintendent are entirely sanguine. From letters received from Mr. WHITE, since the day of his report, an extract from which is also appended, it appears that indications in the further progress of his work fully justified the favorable opinion expressed by him in that paper.

In view of the recent monetary difficulties of the country, from which it has not yet entirely recovered, the operations at the mine have not been pressed forward on as extended a scale as the character of the several veins would seem to render desirable; and the same policy the Board deem to be the prudent course for the present.

Two thousand shares of the stock at the organization of the Company were placed to its credit, which brought $3000 into the Treasury, to meet first expenses. Two calls, one of fifty cents per share, and the other of twenty-five cents, have since been made, as will appear from the report of the Treasurer. Another call of twenty-five cents per share must be made at an early day, which we trust the encouraging prospect of our mining interest will induce stockholders promptly to meet.

Respectfully submitted, by order of the Directors,

E. N. WILLCOX, *Sect'y.*

---

*Extract from Superintendent's Letter under date of June 23d* 1855.

"The new or No. 3 vein, is looking finely, full of copper. We have been successful in uncovering it in several places for 20 rods and more. Have cut a feeder in the adit, with a decided improvement in rock. The feeder carries considerable copper; will send you a piece The rock is finer grained and more settled—better character of trap. I feel more sanguine than ever before.

## STATEMENT OF TREASURER.

---

Amount of Receipts and Disbursements by A. TEN EYCK ESQ. Treasurer, f.om organization of Company to Nov. 23d, 1854, as appears by books of the Company................$11 093,70

Of this amount it appears that he received for stock sold for benefit of the Company, 2000 shares..........................$3,000.00

From assessments,............... 8,003.70

And that he expended the same as follows

For land,..........................$ 448.14

Mining, and all other expenses,......10,645.59

$11,093.70 — $11,093.70

Amounts received and disbursed by E. N. WILLCOX, Treasurer, from November 23d, 1854, to June 15th, 1855.

RECEIPTS.

For stock forfeited and redeemed............................$ 150.00

For assessments................. 2439.61

EXPENDITURES.

For mining expenses paid on drafts of Superintendent of mine.....................$2426.25

Cash on hand,.......................... 163.36

$2589.61 — $2589.61

DETROIT, June 15th, 1855.

E. N. WILLCOX, *Treasurer.*

### STATEMENT OF BOOKS OF CLIFTON MINING CO.

| Capital Stock, being amount received for assessments and sale of stock, viz: | | Profit and loss being amount expended,...................... | $13683.31 |
|---|---|---|---|
| July, 1854, Assessment 25 cents,........ | $5,000 | Assessments due and available... | 1340.19 |
| Sept. " " " " ........ | 5,000 | Stock forfeited,.................. | 2976.56 |
| Jan. 1855, " " " ........ | 5,000 | | |
| 2,000 shares sold by A. Ten Eyck........ | 3,000 | | |
| | $18,000 | | $18,000 |

E. N. WILLCOX, *Treasurer.*

June 15th, 1855.

N.B.—The sum of $1340.19 noted as available in above account, has been received since date of report and expended as per drafts from Supt.

# REPORT OF SUPERINTENDENT.

---

E. N. WILCOX, ESQ., *Treasurer of Clifton Mining Company.*

DEAR SIR: In compliance with your request I submit the following report:

The mining ground of the Company is the S. W. ¼ of Sec. 10 in T. 49 S. of R. 41 W. The first improvements on the ground were commenced in Nov, 1853. Some work of mining and exploring was done during the following winter. The management was then in the hands of Capt. J. D. ANGUS. I entered in charge in May, 1854

The map and vertical section herewith show the face of the mining ground. The first mining was a shaft sunk to the depth of 28 feet. The shaft not proving to be upon a regular vein is not represented. The mining doing when I entered in charge, was sinking on vein No. 2, and driving the adit, both of which were commenced in March. The shaft was continued on the vein to the depth of 35 feet. The vein to this depth has good walls and is from two to ten inches in thickness. It is composed of epidote quartz and spar, with a lining at walls of laumonite. Its course is south about 44 deg. west, and dip northwesterly 40 deg. from the horizon. It carries stamp copper, and bears every feature of a true vein, and I feel sure that at its intersection with the adit, it will prove a *strong* and *rich* vein. The shaft near its present depth is cut by a cross course, which let in much water, and the force in July was taken from it, and placed upon the adit, in order to press the latter forward to effect the drainage.

In June a shaft was commenced on vein No. 4. This vein being at the brow of the bluff, at the surface, was considerably deranged, but at the depth of 20 feet made regularly upon the foot wall and was looking well at the depth of 30 feet, when the work was discontinued in July. Its course was as nearly as could be determined S. 50 deg. W., and its dip 52 deg. northwesterly. It is on the back of vein No. 2, about 200 feet distant and approaches the latter in depth and extension eastwardly, and is the same as vein No. 2 in mineral features.

The extent of the adit in earth is 114 feet. The rock was reached in June, when a shaft was raised to the surface for ventilation.

Resuming on the adit, it has been steadily continued and is now extended 148 feet in rock. Two, four and six miners have been worked about the same time for each number, and the average progress has been about 14 feet per month.

Vein No. 1, was cut out cropping in the face of the rock about 3 feet above bottom of adit. It is 3 feet between walls with a bearing S. 45 deg. W. and dip northwesterly 35 deg. from a horizontal. It has a healthy veinstone of epidote, quartz and spar, and carries occasional specks of copper. Fifty-seven feet from this vein the adit passes through another vein which has the same bearing but a transverse dip, the dip being S. 46 deg. E. This vein, called by miners a floor, is composed of calcareous vein matter, from 12 to 15 inches wide, with a heavy underlaying seam of clay, and has regular and smooth walls. Between the vein and floor, and to a point 10 feet beyond the latter, the rock was very unsettled. It broke and mined favorably however to a point 30 feet beyond the floor. It there changed to hard compact rock, which continued for 50 feet. In this rock the miners could advance but 14 or 15 feet per month. A contract entered upon on 1st of March was 10 feet of it uncompleted at the 1st of June. I had expected to reach vein No. 2 in May, but this unexpected state of ground rendered it impossible. We are now 55 feet from the vein. The rock is becoming softer and breaking freer, and we are expecting it will so continue to the vein.

The object of the adit in addition to drainage of vein No. 2 was to discover any veins, if any, underlaying this vein. The Sharon Mine adjoins our ground on the west. It was earlier commenced and is much more advanced. It has the same position on the mineral range, and the veins which that company have discovered we may look for with certainty. They have opened four different veins—that upon which they have done the most, and with the best promise, which they term their south vein, is near the foot of the south declivity of the range. The surveys and correspondence of features place this vein south of vein No. 2 upon our tract, and within the ground opened by our adit—and the evidences now developed make it very conclusive that *this vein* is *our vein* No. 1. The Sharon Company have sunk upon this vein 110 to 120 feet. Its out crop on their ground is on a higher level than on ours. At the depth of 86 feet they struck this floor or transverse vein, which we have met with in our adit. The features of the floor, as its course, dip, composition, &c., are essentially the same at both mines, and the same is the case with the vein. The vein and floor intersect at our point 29 feet below the adit, which is but 4 feet lower, as determined by level, than their intersection at Sharon. At the level of our adit above the floor the vein at the Sharon carried but little copper and continued poor until the floor was passed, when it rapidly improved and at 15 feet below is 30 inches wide, and very rich in stamp and bbl. work, so much so as to encourage the belief that it would soon make into heavy mass copper. We may reach this point of the vein at 40 to 50 feet below our adit by sinking a perpendicular shaft just north of the floor, and from the entire similarity of the ground, we may expect the vein to prove equally rich.

The last of April I discovered another vein, between veins No. 2 and 4, 140 feet back of the former, which I class as vein No. 3. It is composed of epidote, quartz and spar, is 2½ to 3 feet in width, and very rich in copper. It is much covered with broken rock and earth but has been uncovered at points from 10 to 12 rods. It shows better at surface than any other vein that has been opened in this district. Being limited to a small force, which were working by contract in adit, I have been able to sink upon it only at one point 3 to 4 feet, where it proved to be 2½ feet wide, with good foot and hanging walls and richly charged the full width with lump and stamp copper.

It is now shown that we have four veins, but upon no one has sufficient work been done to prove its value. The discovery by the adit of our south vein is highly important as identifying our position with the same portion of the mining range as that of the Sharon Co., and the proof of this vein, which this company have made, is highly encouraging to us.

In the farther carrying on of the work, I would recommend to sink a shaft in the adit to strike vein No. 1, below the floor and open upon this vein, extend the adit, which is the best way of proving the ground when there is so much surface soil, and to make proof of vein No. 3 by sinking a shaft upon it. All this work is very desirable, but if more than this is at present deemed advisable, I would then recommend as the immediate work, the sinking of the shaft on vein No. 3.

The amount of shafting at present done is 123 feet, and amount of drifting 262 feet.

Our surface improvements are so advanced that this expense will be very light for the coming year. We have 10 acres cleared and in crops, 2½ acres are potatoes, one acre turnips, and the remainder in oats. Our products last year were 300 bushels potatoes, 20 bushels turnips, and 3 tons oats—from 5 acres improved. Our stock consists of one span horses, two hogs, and 12 pigs. Our buildings are a boarding-house 24 by 26 feet, a ware-house 22 by 24 feet, an office 18 by 22 feet, a blacksmith-shop 18 by 24 feet, and four out-houses. We have on hand 160 cords of wood, with enough fitted for the stove to last to another winter; 4000 feet lumber, 3½ M. shingles, 400 feet hewn timber and saw-logs for feet lumber.

The lumber, shingles, &c., were got out by Capt. ANGUS for another house, but which we have not seen fit to erect. Our building accommodations are now adequate for a force of 25 men. Further building, or surface improvements will not be required until the mining is much more advanced. The land improved is sufficient to raise all products required, with the exception of hay, and this, should it be wanted, can be procured at favorable rates from a meadow at a short distance, owned by myself. The following are accounts of expenditures, assets and liabilities to June 1st, 1855:

| | | |
|---|---|---|
| Total expenditures for acc't of mine to June 1st, '55 | | $12701 44 |
| Mining, including mining labor and material, windlassing, wheeling and cross-cutting | $3241 78 | |
| Surface, including clearing land and raising crops | 1822 02 | |
| Teaming | 168 38 | |
| Building | 1115 83 | |
| Boarding-house, including supplies | 2955 24 | |
| General expenses, including salary of agent, traveling, stationery, and postage | 1222 88 | |
| Freight | 992 73 | |
| Blacksmith, including labor, tools, material | 702 27 | |
| Roads | 422 84 | |
| Store | 57 47 | |
| | | $12701 44 |
| Assets at mine, Building and surface improvements | 2450 00 | |
| Horses, harnesses, wagon, sleighs, hogs, farming implements, carpenters' tools, mining tools and lumber | 1341 00 | |
| Supplies, bedding, furniture, materials, and office stationery | 715 24 | |
| Bills receivable | 226 87 | |
| Cash on hand | 38 85 | |
| | | $4812 96 |
| Liabilities at mine June 1st, 1855 | | |
| Bal. due employees | $1711 00 | |
| Freight and storage | 77 57 | |
| Bal. being surplus of effects to this date | 3024 29 | |
| | | $4812 96 |

Respectfully yours, F. G. WHITE, *Agent.*

Clifton Mine, June 1st, 1855.

# ARTICLES OF ASSOCIATION.

Article 1. This Company is organized under an Act of the Legislature of the State of Michigan, entitled, "An Act to authorize the formation of Corporations for Mining, Smelting or Manufacturing Iron, Copper, Mineral Coal, Silver, or other Ores or Minerals, and for other Manufacturing purposes;" approved February 5th, 1853. Its name shall be the Clifton Mining Company, and its Capital Stock shall be Five Hundred Thousand Dollars, divided into twenty thousand shares of twenty-five dollars each.

Art. 2. The business of said Company shall be the Mining, Smelting and Refining of Copper and Copper Ore and other Metals and Metallic Ores; and its business shall be conducted in Ontonagon Co., in the State of Michigan, and its Business Office kept in the city of Detroit.

Art. 3. The Officers of said Company shall consist of seven Directors, a President, Treasurer, and Secretary, and such other Agents as the Directors may from time to time appoint. The President or Secretary may be also Treasurer.

Art. 4. The annual meeting shall be held at the office of the Company in Detroit, on the second Tuesday of June, and may be adjourned from time to time. The time and place of holding the annual meeting may be altered by the By-Laws.

Art. 5. The amount paid in on the Capital of said Company is thirty thousand dollars. The Stock of said Company is held by the following persons, in the following amounts: Charles Richmond, of Detroit, Wayne County, Michigan, holds four thousand Shares; John R. Grout, of Detroit, Wayne County, Michigan, holds four thousand Shares; Henry N. Walker, of Detroit, Wayne County, Michigan, holds four thousand Shares; Charles P. Woodruff, of Detroit. Wayne County, Michigan, holds four thousand Shares; Algeron Merryweather, of Pontiac, Oakland County, Michigan, holds four thousand Shares.

Art, 6. The said Company shall continue in existence thirty years. In witness whereof we have hereunto, and to two instruments of like tenor and date, set our hand this first day of October, in the year Eighteen Hundred and Fifty-Three.

Signed,

JOHN R. GROUT,
CHARLES RICHMOND,
C. P. WOODRUFF,
HENRY N. WALKER,
A. MERRYWEATHER.

STATE OF MICHIGAN } *ss.*
County of Wayne. }

On this first day of October, in the year Eighteen Hundred and Fifty-Three before me, a Notary Public, in and for said County, personally appeared the above named Charles Richmond, John R. Grout, Henry N. Walker, Chaales P. Woodruff, and Algernon Merryweather, known to me to be the persons described in, and who executed the foregoing articles, and acknowledged that they had executed the same for the uses and purposes therein expressed.

Signed, JAMES V. CAMPBELL,
*Notary Public, Wayne County, Mich.*

I do hereby certify the above and foregoing to be a true copy of the Articles of the "Clifton Mining Company," and of the Certificate of Acknowledgment thereunto attached received and filed in the Office of the Secretary of State, October sixth, A.D., 1853.

{ L. S. } In testimony whereof I have hereunto set my Hand, and affixed the great Seal of the State of Michigan, at Lansing, this twelfth day of October, A. D., 1853.

ROD. P. GIBSON,
*Dep. Secretary of State.*

# BY-LAWS.

ARTICLE. 1. Each Stockholder shall be entitled to receive Certificates of the Shares of Stock held by him, signed by the President and Secretary.

ART. 2. Twenty days notice of all assessments shall be transmitted by mail to each registered Shareholder, and notice of each assessment shall be published in a Detroit daily newspaper, at least twenty days previous to the time of payment.

ART. 3. In case any Stockholder shall neglect, or refuse payment of any assessment, for the space of sixty days after the same shall have become due and payable, and after he shall have been notified thereof as in the By-Laws provided, the Stock of such delinquent Stockholder may be sold at public auction, at the office of the Secretary, upon thirty days' notice in some newspaper published in the County where the office of the Secretary is located, and the proceeds of such sale shall be first applied in payment of the assessment due and the expenses on the same, and the residue shall be refunded to the owner thereof; and such sale shall entitle the purchaser to all the rights of a Stockholder, to the extent of the Shares so purchased.

ART. 4. The Directors may, by Resolution, accept the surrender of Shares of Stock and relieve the holders thereof from all assessments, and the Shares so surrendered may be disposed of as they shall determine.

ART. 5. Stocks shall be transferable upon the books of the Company only, and it shall not be competent for any transfer to be made while any assessment remains unpaid.

ART. 6. Regular meetings of the Stockholders shall be held annually on the second Tuesday of June in each year, and special meetings at such other times as the Directors may appoint. The Stockholders may appear in person or by proxy duly filed with the Secretary, and be entitled to one vote for every Share of Stock held by them. The Directors shall, on the written request of any person or persons representing twelve hundred Shares, call a special meeting. All meetings called by the Directors, shall be held at the office of the Company in Detroit, and notice thereof transmitted by mail to each registered Shareholder, and published in a Detroit daily newspaper at least twenty days before the time fixed for the same.

ART. 7. Meetings of the Directors may be held as often and at such time and place as they may determine, and at any such meeting a majority of their body shall constitute a quorum for the transaction of business, and the President, or, in his absence from Detroit, any Director may call the same.

ART. 8. The Directors may, from time to time, direct the compensation of all officers of the company.

ART. 9. The Secretary of the Company shall keep full records of all the proceedings of meetings, both of the Shareholders and Directors, and notify and publish all notices required by the By-Laws, preserve and file all papers pertaining to the business of the Company, issue all certificates contemplated by the By-Laws, and perform all such other duties as appropriately belong to the Secretary. He shall be entitled to such annual compensation as the Directors may determine.

ART. 10. The Treasurer of the Company shall receive and safely keep all moneys belonging to the Company, and pay the same out as required by the Directors. He shall keep true accounts of the finances of the Company, subject at all times to the examination and inspection of the Directors and Committee of Stockholders appointed for that purpose. He shall execute a bond to the Company, with sureties to be approved by the Directors in the penal sum of          thousand dollars, conditional for the faithful discharge of the duties of his office, and he shall be entitled to a salary to be fixed annually.

ART. 11. The Directors may declare dividends of the profits of the corporation, when and as often, as they may deem that the state of the funds will permit, giving twenty days public notice, in the same manner as for call of assessments, of their having done so, and of the time and place of payment of every such dividend—such payment to be made to those Stockholders who shall appear on the Register ten days prior to the day of such payment; the transfer books to be closed during said ten days.

ART. 12. The Directors for the time being shall have power to fill any vacancy which may happen in their board, by death, resignation, or otherwise, for the current year, and they may remove any officer, agent, or servant of the Company for good cause.

www.ingramcontent.com/pod-product-compliance
Lightning Source LLC
LaVergne TN
LVHW020636110826
845149LV00004B/1220